William Anderson

The gold fields of the world

Our knowledge of them and its application to the gold fields of Canada

William Anderson

The gold fields of the world
Our knowledge of them and its application to the gold fields of Canada

ISBN/EAN: 9783337257835

Printed in Europe, USA, Canada, Australia, Japan

Cover: Foto ©berggeist007 / pixelio.de

More available books at **www.hansebooks.com**

THE

GOLD FIELDS OF THE WORLD

OUR KNOWLEDGE OF THEM

AND

ITS APPLICATION TO THE

GOLD FIELDS OF CANADA

A COMPILATION

BY W. J. ANDERSON

M. R. C. S. E.

QUEBEC:

PRINTED BY G. & G. E. DESBARATS.

1864.

THE

GOLD FIELDS OF THE WORLD,

&c.,   &c.,   &c.

It has been said by them of old, " the love of gold
is the root of all evil," and the Bard of Avon has
sung:

> How quickly nature falls into revolt,
> Where gold becomes the object :
> For this the foolish overcareful fathers
> Have broke their sleep with thoughts, their brains with care,
> Their bones with industry,

And the finale

> Are murdered for our pains.

But, however, much the inordinate desire for gold
and its abuse, may have been denounced by poets and
moralists, the common sense of mankind has arrived
at the conviction, that gold has been an instrument
of much positive good, and that its acquirement and
increase by individuals and communities, are not
only legitimate but praiseworthy, as calculated to
promote individual comfort and happiness, as well as
the order, peace, prosperity and civilization of nations.
And of late the discovery and opening up of gold
fields has been the study of philosophers and statesmen.

As very vague and erroneous views have been
prevalent on the subject, it is proposed to give here
briefly, the natural history of gold, and our present
knowledge of the gold fields of the world, and to
show how that knowledge may be practically applied
to develop the gold fields of Canada.

DISTRIBUTION OF GOLD.

Gold has been almost universally distributed over the Globe. From the remotest antiquity, it has been known and esteemed one of the most precious metals. It is probable that the property which contributes most to its intrinsic value, viz., its *indestructibility,* and its being found *native,* or in a pure metallic state, almost on the surface, led to its early discovery.

Gold is *invisibly* diffused through the soil, from which it is eliminated by plants, and from this source there was procured from the ashes of grape vines, grown on the Seine, as much gold as made several Napoleons. Gold is found scattered through alluvial and diluvial deposits, in *spangles,* one thousand of which will not weigh a grain, and in *pepites* or " nuggets " of every variety in form and weight. The Australian nugget in possession of Dr. Kerr, weighs 106 lbs. troy of pure gold. The Pepite deposited in the Russian School of Mines, weighs 97 lbs. The Californian nugget 28 lbs.

There is also " a mine for the silver, and a bed for the gold which men refine."

Gold occurs in nature in the following forms :

1. Alloy of Gold and Silver, with traces of Iron, Copper and other Metals.

2. Porposite—Gold and Palladium.

3. Rhodium Gold—Gold and Rhodium.

4. Gold Amalgam—Gold and Mercury, with a little silver.

5. Graphic Tellurium ore—Telluride of Silver and Gold.

6. Aurotellurite—Telluride and Antimonide of Gold, Silver and Lead.

The *first* combination furnishes almost the whole of the gold in the world, and till lately nine tenths of it were procured by washing diluvial or alluvial deposits.

Native gold is *invariably* found alloyed with silver, and there may be also traces of other metals, especially iron and copper. It is therefore said to be of varied fineness, to denote which it has been arbitrarily divided into twenty-four parts, or *carats*, each carat being four grs. Refined gold is of twenty-four carats fineness, which means that the ninety-six grs. are all gold. British coins are of the fineness of twenty-two carats, that is to say, every ninety-six parts contain eighty-eight of gold and eight of alloy. The lowest stamped gold plate is 18 carats fine, being three fourths gold and one fourth alloy.

FINENESS OF GOLD.

	Gold.	Silver.	Copper.	Iron.
Australian	99.283	.437	.069	.20.3
Russian	98. 96	. 16	. 35	. 5.
Georgian	95.579	4.421	trace	trace
Brazilian	94. 00	5. 85	"	"
California	90. 00			
Canada	86. 40	13.60	"	"
N. Carolina	65. 03	34.18		

The Gold of California is remarkably constant in fineness, yielding 90° ; Australian gold varies 12° ; and Russian 18° ; one gold region and even different claims in the same gold field, varying from each other. Dr. Hunt has given 86.40 as the standard of specimens tested, but samples of gold procured last season from the Chaudière district, indicate from their appearance, a fineness equal to that of Australia.

But in whatever form or under whatever conditions we find gold, its source is one and the same. " The earth poureth forth bread, *but below it windeth a fiery region ; sapphires are its stones, and gold is its ground,*" and from this mighty laboratory of nature *has* proceeded, and still *is* proceeding, the metallic wealth of every region.

Till some time after the discoveries in California and Australia, the chief supply of gold was from the diluvial and alluvial *drifts*, and the opinion most generally entertained was, that it originated exclusively in the Palæozoic strata, in the Lower Silurian system, and Murchison had undertaken to shew that the impregnation of the rocks of the Ural with gold, took place at a very recent geological period, as late as the drift epoch, and he gave utterance to his well known anxiom, " That it would be vain to assign limits to the productiveness of silver mines, when science had been fully applied to them, *for they increase in value as in depth*, whereas *gold diminishes as we descend to seek it.*" But recent and more extended experience has modified Sir Roderick's views.

Gold there is reason to believe is equally diffused, though invisibly, through the lowest or most ancient rocks of the Azoic period, though geologists were first aware of its existence in those lower strata of the so-called Palæozoic series, which have undergone a *metamorphosis* or change of structure, by igneous agency, or some other power. These strata having been upheaved by volcanic, or some slower action, form the great mountain ranges, which by abrasion and disentegration during the glacial and diluvial periods, were the sources of supply of gold to the deposits of those epochs, and have continued to be the chief sources to the present day.

The general facts in regard to the mineralogical characters of the gold bearing rocks, will be shewn to be very nearly the same all the world over. Whatever was their original structure or composition, they have by the agency of a long train of geological events, been brought to exhibit a striking resemblance to each other. They consist most frequently of slaty rocks, more generally talcose, although occasionally chloritic or argillaceous ; and it is in these rocks, that the gold bearing quartz, which almost invariably forms the *gangue,* or accompanying mineral is found to be most productive.

But gold is not confined to these formations, for it is sometimes found in remunerative quantity, in the hypogene, eruptive, or intrusive rocks, a notable example of which is the vein forming the mine of St. John del Rey in Brazil, a full description of which will be hereafter given ; indeed all the phenomena in connexion with gold throughout the world, will be found to prove, that gold has not been confined to any particular geological period, on the contrary, there has been, and there is still going on, a repetition of similar conditions from the earliest period to the present time.

Gold is found in superficial, stratified and unstratified deposits.

SUPERFICIAL DEPOSITS, comprise all which lie loose on the surface, or are intermingled with the superficial formations, the drift, and the alluvial, and are made up of particles and rounded fragments, which originally formed a part of some bed or vein. Such deposits are termed " placers " or " washings."

STRATIFIED MINERAL DEPOSITS, comprehend those masses or veins which are *included* within rocks of sedimentary origin, or the metal may be disseminated through them so thoroughly, as to be undistinguishable.

UNSTRATIFIED DEPOSITS, may be *regular* or *irregular.* IRREGULAR deposits, include igneous eruptive rocks, and are found most largely developed in the Azoic system, having been poured out when the crust of the earth was thinnest and most easily fractured. There have also been eruptions at later periods, occurring in volcanic regions, or on deeply seated fissures, which have remained open, until a late epoch. These formations being developed on a very large scale in connexion with the baser metals, more especially iron, have been a long time known and worked. But there are allied with these, certain eruptive rocks, in which the metal does not form a special deposit, such are the " *traps* " and " *serpentines* " in which gold and platinum and associated metals are disseminated in fine particles ; such also is the *jacotinga* of Brazil, which consists mainly of specular iron, mixed with sulphuret of iron, magnetic pyrites and quartz, through which gold is equally diffused. REGULAR deposits are veins which may be either *segregated*, *gash*, or *fissure*. Whitney defines a vein to be " an aggregation of mineral matter, of indefinite length and breadth, and comparatively small thickness, *differing in character from, and posterior in formation to*, the rocks which inclose it."

Plutonic veins, are fissures filled with matter which has been forced in a plastic state from *below* upwards.

Sedimentary veins, are fissures filled from *above*, by deposition of mineral matter, in a manner similar to that in which the sedimentary rocks have been formed.

Stalactic veins, or veins of infiltration, result from the filling of fissures, by incrustation of the sides with calcareous matter deposited from aqueous solution.

The results of all these actions may be exhibited in the same fissure.

Segregated veins are gangues, which do not seem to occupy a previously existing fissure, but in which the metallic and mineral substances have been gradually eliminated from the surrounding formations, most probably by electro-magnetic action, or by chemical. They usually lie *parallel with the cleavage planes* of the formation in which they occur, the gneissic and schistose portion of the metamorphic rocks. But its extent *downwards is not to be relied on.*

Gash veins, like segregated, occupy preexisting fissures, but of limited· extent. They are usually confined to a single member of the formation in which they occur, and terminate below, *when a change in the character of the rock takes place.* The origin of these veins is referred to contraction of the rock, caused by shrinkage, either while undergoing consolidation, or by fracture of certain strata on cooling.

True veins, are fissures in the solid crust of the earth, of indefinite length or breadth, which have been filled with mineral matter. They have originated in " faults " or dislocations caused by great dynamical agencies, connected with extensive movement of the earth's crust, and thus are believed *to extend indefinitely downwards :* no well developed vein having yet being found terminating *in depth.*

Admitting these definitions to be correct, there can be no difficulty in reconciling the discrepancies of opinion which have existed on Sir Roderick Murchison's axiom on gold deposits. It is doubtful whether experience has yet furnished the means by which we can, a priori, give the character of any vein when first discovered, but we now know this to be the fact, that should it prove either a segregated or gash vein, its product of gold will be uncertain and diminish, and ultimately cease, as we descend. On the other hand, should it turn out to be a true fissure vein, all the probabilities are in favor of its increasing in richness as in depth.

GENERAL DESCRIPTION OF GOLD REGIONS.

We will now proceed to illustrate our subject by a condensed account of the principal gold fields of the world ancient and modern.

SOUTHERN ASIA.—The continent south of the great Himalayan chain in former times yielded gold in great quantities, and it is still found in many localities both on the continent and the islands of India, China and Japan, though in diminished quantity. It was obtained from washings. From India, Solomon is supposed to have procured $1,500,000, and from Ophir, wherever it may have been, he brought over a million more.—The washings of the Burrampooter are still estimated to yield annually, about 3,000 lbs. troy.

AFRICA.—The golden sands of Africa and its quartz mines yielded in ancient times, a vast amount of very fine gold. Agatharchides who wrote, one hundred years before the Christian era, describes the mode of mining and amalgamation (very similar to that of the present day) which was followed by the Egyptians in working the mines of the quartz bearing rocks of the Red Sea, from the most remote periods. These mines were long ago abandoned ; but copper tools and skeletons of the wretched beings who were compelled to work them, are still found in the galleries and passages. Modern attempts to mine, by the Pasha of Egypt, and by the aid of Europeans, have proved very disastrous. The climate alone being a sufficient hindrance. But the sands of Africa still furnish annually about 4000 lbs. of gold.

GREAT BRITAIN.—In South Wales, the Romans mined extensively in the Silurian rocks, with but little return of gold. In Scotland, several hundred men were employed in washing gold from the sands in Dumfrieshire, in the reign of James V. For hundreds of years it was observed in the streams of Cornwall and Devon, and a few ounces are still obtained yearly by the miners. One piece weighed 15 dwts. 3 qrs. In Ireland, the Wicklow Hills are of the Geological type which furnishes gold, and one piece was found weighing 22 oz. Much excitement followed, and a large capital was wasted in the endeavour to find the *lode* from whence it came. The yield of Great Britain for a few years previous to 1853, was about 4 oz. per annum, but after the publication of John Calvert's " Gold Rocks of Great Britain," an impulse was given to quartz crushing, and many samples of quartz and gossan have yielded several ounces, and some as high as sixteen ounces of gold to the ton. It would appear however that the lodes as yet discovered are not very wide and do not extend to a depth exceeding twenty fathoms. Mr. John Taylor, junr., whose experience in these matters has been very great, says, " I have seen evidence to make me believe, that British gold ores, in moderate quantities can be obtained, and that if they are skilfully treated, they can be *made to yield a moderate profit*, but beyond this I cannot persuade myself that producing gold in England, will either be a large or very lucrative branch of industry."

FRANCE.—The Gallic rivers were once rich in golden sands, and the sands of the Rhine are still washed on a small scale. The washer makes about seven francs per day. The washings are now very poor, as will appear from the following comparison : Rhine, 1; Siberia, 10 ; Chili, 37. Only one auriferous quartz vein is known in France, it is in the department of the Isére, it was discovered in 1700 and worked at intervals up to 1841, yielding a small amount. Many of the lead veins contain a little gold.

GERMANY.—Here the amount of gold is very small, though in some localities, mining and washing have been uninterruptedly pursued from the time of the Romans.

In HUNGARY, gold, silver and copper mines have been worked steadily since the eighth century, and the result of these long continued operations present a most interesting field of investigation. Here are to be found ancient works on the most gigantic scale. The ores are chiefly auro-argentiferous ; galena is also obtained in sufficient quantity to afford lead for separating the silver. These mines are much less productive than formerly, but works are in progress on a gigantic scale *to open them at a great depth,* though the quantity of the *precious metals* has been found to decrease, as the veins are worked downwards. The rare and curious combinations of gold and tellurium are found in Transylvania. In 1854 the annual product was 2850 lbs. gold and 57,000 lbs. silver. The Hungarian mines could not be worked profitably unless under the highest development of metallurgic and mineral skill, and the lowest price of labor.

The same may be said of the other portions of the Austrian dominions, but from skill and economy in management the total product of the Empire, increased from 2,682 lbs. in 1820, to 5,645 lbs. in 1848, and 6,000 lbs. in 1854.

RUSSIA.—Russian gold is obtained almost entirely from the eastern slope of the Ural mountains, from Siberia and the Caucasus. It was first discovered in 1743 near Ekatherinesberg, *and the first workings were in the solid rock,* in 1752 at Bresow, they are still productive though in a diminished degree. In 1823 gold had been mined from the solid rock in sixty-six localities, but with the exception of eight, all had been abandoned. *The veins are numerous and contained in* GRANITE, *which itself forms veins in the talcose, chloritic and micacious slates.* The productive ores are of quartz, *and cut the granite masses at right angles, having nearly a vertical dip.* They do not generally extend beyond *the limits of the granite,* and the workings demonstrated that the amount of gold decreases according to depth. No one on reading this description can doubt that these are either *seg-*

regáted or *gash* veins, and the results should not surprise.

The proper gold washings of the Ural which have produced such large amounts of gold, and which were so famous before the discoveries in California and Australia, *were commenced by the Crown* in 1814, from which date to 1850 they yielded 803,700 lbs., while the rock yielded 128,570 ; total 932,270.

The maximum was reached in 1847, since when there has been a decided diminution, and Whitney says, " It may be safely inferred that it will continue slowly to decline."

The washing machinery of the Ural is remarkably efficacious, much time and money having been expended in its perfection. It would require to be so, as the washings only yield ¾ *of a grain of gold, to the bushel of* 100 *lbs.*

SOUTH AMERICA.—In the year 1800, Brazil from its washings, procured 9,900 lbs. of gold ; it had been famous during the whole of the 18th century as a gold producing country, but since 1800 we have precise information, and we know that the washings are nearly exhausted, and that the principal source of supply is the veins in the solid rock. The Brazilian auriferous rocks *differ* from those in other parts of the world. The rocks, if Palæozoic, are so metamorphosed, as not to be referable to any precise epoch. They have not been elevated to Cordilleras, but form a series of dome like protuberances and dykes, rather than mountain chains.

But there can be no doubt that the " itabarite" or jacotinga already alluded to, is an igneous eruptive rock forming one of the largest true fissure veins in the world.

As the Brazilian mines *in the solid rock*, are the only ones which up to 1854 had been worked exten-
2·

sively and profitably, it is important to have them correctly described.

Whitney says, the gold is found in jacotinga, intercolated with argillaceous itacolumite slate of a *reddish blue* color, the gold being equally and invisibly diffused through it. The auriferous mass averages about 44 feet in width. It dips with the rocks of the vicinity at an angle of 45° to the south-east.

There are three mines known as Bahu, Cachoeira and Gambu. The principal shafts are sunk, inclining about 45° with the auriferous bed, and the ore is raised on a tram-road by cars. The Bahu mine in 1854, was 1,200 feet deep. As the ore comes up it is broken by *negro women*, and carried to the stamps. The coarse gold is caught on cowhides with the hair on, which are changed every two hours, while the slime is amalgamated in barrels. In each barrel are placed 80 lbs. of mercury to 16 cubic feet of slime, and the whole is allowed to revolve for thirty hours. The loss of mercury is 728 lbs. or 0.037 lbs. per cubic foot of slime, amalgamated. The yield of gold, about $\frac{5}{8}$ oz. per ton. The gold contained 20° silver.

These mines had been worked one hundred years previous to 1834, and had been abandoned as exhausted. The St. John del Rey Co. to whom they belong, commenced operations in that year, but did not work them profitably till 1858, since which time they have been steadily increasing in value. In 1846, they crushed 34,935 tons of rock, which yielded 1,465 lbs. troy of gold ; in 1852, they crushed 82,642 tons, yielding 3,233 lbs. gold, being about 9 dwts. 9 grs. per ton.

The net profit on the fiscal year 1852 and 3 was £55,390 16s. sterling. The total profits from 1838 to 1852 being £312,621 sterling. These mines deserve attention, first, on account of the great size of the vein ; second, on account of the great depth to which

it has been worked without diminution of the product
of gold ; and lastly, as shewing how an ore poor in
gold, but abundant in quantity, when worked with
economy and skill may become a most profitable investment.

There are other mines in New Grenada and Vene-
zuela which are also being worked advantageously
by English companies. The greatest yield was about
the middle of the 18th century. From 1752 to 1861,
the royal quint, was paid on from 17,000 to 21,500 lbs.
annually. In 1822 the yield, from the failure of the
washings, fell off to less than 1,000 lbs.

At present the English Companies mining in the
rock, furnish almost all the gold. *The washing has
nearly ceased*, but the annual product had been raised
in 1854 to 6,000 lbs.

NORTH AMERICA.

There are two great gold regions in North America,
one on the Atlantic slope, the other that of California.
A lump of gold weighing 17 dwts. was first found near
the Rappahannock in Virginia. In 1799, a lump of
gold about the size of " a small smoothing iron" was
found in N. Carolina. Up to 1825 gold was only
procured from washings, but at that date, valuable
quarts veins were discovered, and though rudely
worked the product was so great as to excite general
attention.

In 1829, Mitchell published a map of the gold region
of N. Carolina shewing nine different mining localities, three being in the Primary and six on the " tran-
sition " or slate. Then followed discoveries in S.
Carolina and Georgia, and as discovery after discovery
was made the limits of the gold region were extended
to Canada. These however all sank into insignifi-
cance, when in 1847, the discovery of gold in Califor-

nia was announced, and the rush of gold seekers that immediately took place, was unexampled in the history of the world. St. Francisco was not only emptied but the tide flowed from Mexico, Peru and Chili; from China and Australia, and lastly from the U. States and Europe. In November of the first year four thousand persons were at the diggings, and the gold taken out was valued at $5,000,000. So rich were the washings that it was not uncommon for a single miner to obtain ten ounces in a single day. In 1849 about 55,000 miners were at the diggings, and the gold obtained amounted to $40,000,000, which was increased in 1850 to $50,000,000.

Whitney, writing in 1854, regrets that there had been no proper Geological survey, and expresses the opinion that from the facts then known, much light could not be thrown on the probable future of the country. It was however known that the Sierra Nevada, was made up of granite rocks, through which volcanic fires had occasionally found vent, and piled up lofty masses of debris, these were flanked by slaty rocks in which the talcose predominated; these alternated with trappean and serpentine masses, which extended to the valley of the Sacramento, where they were, concealed by sedimentary deposits of recent origin. The slates of the Sierra, formed a belt of fifty miles wide extending through the valley; they are very ancient, and if they ever belonged to the Palæozoic series, all traces of organic remains have been obliterated, and they have assumed a crystalline structure, being every where invaded and broken up by igneous masses. They may be classed with the lower Silurian of the Andes and the Ural, and are undoubtedly the source of the gold of the placers.

Up to 1854 almost the whole gold had been obtained from the *superficial deposits*. At the close of 1853, there had been deposited at the Mint, gold to the value of $224,392,334, of which $207,316,177, were the product of California. The maximum was raised in 1851, 262,000 lbs; and there was a gradual decline to 1854, when the yield had fallen to 252,000 lbs, valued at $62,500,000.

When the placers began to shew signs of exhaustion, attention was turned to quartz mining, and in 1853 at least twenty Anglo Californian Cos. for quartz mining were formed in London, most of them having no existence except on paper, but in January, 1859, there were 3,000 mills in operation, with an aggregate of 2610 batteries; and the cost of machinery was estimated at $3,270,000.

The total amount of gold produced in the whole United States in 1862, according to Mint returns, was $30,976,593 24 cents and $1,032,264 45 cents silver; against $55,622,051 gold in 1853.

The total amount of gold of California received at the Mints up to June 30, 1862, was $528,145,665, and the entire yield was estimated at $650,000,000; and within the same period $3,000,000 of silver had been extracted from the gold.

The effect on the progress of the country, was the increase of population to 379,994; flour exported in 1861 to the value of $3,583,700; leather a large article of commerce; and the successful introduction of the tea and coffee plants.

We cannot conclude the reference to the California gold fields better than by a few extracts from the "National Almanac" for 1863, and "Hittel's Resources of California."

There are "says the Almanac" three distinct gold regions, though the first two are connected by outlying placers and leads.

1st. Eastern range containing 1,000 square miles of available mining territory, quartz veins and placers;

2nd. Middle placers, 6,000 square miles, mainly gold washing ;

3rd. Valley mines extending from north to south a linear distance of 250 miles equal to 6,000 square miles.

Hittel says : " The rich quartz veins of California extend from Kern River to the Siskiyou, are found on hills, in canons and in vales. They are at least two thousand feet above the level of the sea, and not more than ten thousand above it. Their course is generally from N. N. West to S. S. East, and they dip steeply to the Eastward, being sometimes nearly perpendicular. They differ in thickness from a line to sixty feet. Quartz veins are very numerous in most of the mining districts, so the task is not to find the veins, but rather to find those which are auriferous.

" Experience has not proved whether large or small veins are more likely to contain gold. It is found in both. The porous quartz, or that containing many cavities, is more frequently found auriferous than the very compact quartz. The best gold-bearing quartz veins are usually yellowish or brownish in tinge, near the surface at least, but very rich specimens are found in whitish or blueish white rock. Most quartz veins in California contain a little gold ; the metal seems to have been distributed most lavishly, but unfortunately in nine-tenths of the veins the proportion of metal is too small to pay. Most of the large veins are supposed to run for miles, though they can rarely be traced on the surface clearly, for more than a furlong. The auriferous veins vary much in thickness. No vein is wrought for more than a few hundred feet ; beyond that it is either too poor to pay or the vein is hidden. Sometimes a vein seems to spread out, and divide into a number of smaller veins, all of which afterwards unite again. These points of junction and the narrower portions of it, are usually richer than other parts of it. When two veins cross each other, one vein may be auriferous on one side of the intersection and not on the other ; but in this case the other vein will be auriferous on both sides.

" Lodes lying between two different kinds of rocks, are usually richer than those which have the same kinds of rocks on both sides. Thus it is said that the richest veins of auriferous quartz in California have been discovered at the intersection of trap and serpen tine, and the richest places in veins, are where they cross from one kind of bed rock into another. The richest part of a lode of auriferous quartz, is almost invariably on the lower side of the vein, near the foot wall. All these are facts to be remembered by the prospector, as a guide and assistance to him, in his search for rich gold-bearing veins.

" If the lode is covered with earthy matter, he may sometimes trace its course, by the difference in the color of the dirt and stone over it, from that elsewhere. When the prospector finds dirt and stones on a vein, evidently disintegrated portions of it, he should wash some of the dirt in a pan, and if he finds no gold, the presumption is that the vein is barren."

So much for quartz lodes, and now here is his description of " Placer Mines."

" Placer mines are divided into many classifications. The first and most important, into deep and shallow. In the former the pay dirt is found deep, twenty feet or more below the surface ; in the latter, near the surface. The shallow or surface diggings are chiefly found in the beds of ravines and gullies, in the bars of rivers and in shallow flats. The pay dirt is usually covered by layers of barren dirt, which is sometimes washed and sometimes left undisturbed, while the pay dirt is taken out underneath it by tunnels or shafts."

In California the placer diggings are generally near the surface ; in Australia, generally deep. After div-

iding into deep and shallow, the next classification is
according to their topographical position, as into hill,
flat, bench, bar, river-bed, ancient river-bed, and
gulch mines.

" Hill diggings are those where the pay dirt is in
or under a hill. Flat diggings are in a flat. Bench
diggings are in a ' bench' or narrow table on the side
of a hill above a river. Benches of this kind are not
uncommon in California, and they often indicate the
place where the stream ran, in some very remote age.

" Bars are low collections of sand and gravel, at the
side of a river, and above its surface at low water.

" River bed claims are those beneath the surface of
the river at low water ; and access is attained to them
only, by removing water from the bed, by flumes or
ditches.

" Ancient river bed claims are those of which the
gold was deposited by streams, in places where no
stream now exists.

" Gulch claims are those in gullies, which have no
water, save during a small part of the year."

In reference to the celebrated " blue lead," Mr.
Hittel says:—" This is not one of the many petty
leads an inch or two in breadth or thickness, which,
after being traversed a few hundred feet, end as sud-
denly and mysteriously as they commence ; but it is
evidently the bed of some ancient river. It is often
hundreds of feet in width, and extends for miles and
miles, a thousand feet below the summits of high
mountains, and entirely through them. Now it crops
out where the deep channels of the rivers and ravines
of the present day have cut it asunder ; and then,
hidden beneath the rocks and strata above it, it only
emerges again miles and miles away. Wherever its
continuity has been destroyed, the river or gulch
which has washed a portion of it away, was found to
be immensely rich for some distance below, and the

materials of which the bed is composed are found with the gold in the bed of the stream. It is evidently the bed of some ancient stream, because it is walled in by steep banks of hard bed rock, precisely like the banks of rivers and ravines in which water now runs ; and because it is composed of clay, which is evidently a sedimentary deposit, and of pebbles of black and white quartz, which could only be rounded and polished as they are by the long continued action of swiftly running water. The bed rock in the bottom of this lead has its roughnesses and crevices, like other river beds. The lighter and poorer qualities of gold are found near its edges, while the heavier and finer portions have found their way to the deeper places near the centre. Trees and pieces of wood, more or less petrified and changed in their nature, which once floated on its waters, are also everywhere encountered, throughout the stratum."

In addition to gold, this lead contains in great quantity, arsenical and iron pyrites. Gold worth millions of dollars has been taken from it, and its richness in the portions longest worked is yet undiminished. The tunnels for working it have cost from $80,000 to $100,000 each.

COLORADO,—Gold discovered in 1858, not remunerative ; but in 1859 rich placers were discovered on the Platte, near Demur City and immense immigration and great suffering and want followed. *The Placer diggings soon gave out*, but quartz veins were found charged with *Iron pyrites* and it was thought that quartz mining would be unprofitable, but in 1861, it was discovered that the pyrites *was very rich in gold*, and mills that had been thrown away came into play. It is now found that these lodes *grow richer as they are opened to a greater depth ;* and in 1862 they yielded very rich returns.

Governor Evans reported that the " Gregory " diggings would yield $5,000,000, and the others, at least as much more ; and that their productiveness would be only limited by the number of miners employed. The country is little suited for agriculture, being chiefly part of the great American desert, yet the population had increased from 42,538 in 1860 to 70,000 in 1862.

NOVADA.—This whole territory is rich in gold and silver. Storey County sent a contribution of $20,000 to the Sanitary Commission in 1862, in eight bars of 111 lbs. each.

OREGON.—These mines in 1861 yielded richer returns than any in California. Estimated product that year $12,000,000.

AUSTRALIA.

The schistose rocks of the Australian Cordilleras, Micacious, Argillacious and Silicious slates, interlaminated with granite have a general north and south strike and stand nearly vertical, they are supposed to be of the Silurian age, but like those of California and Ural have been highly metamorphosed, and broken up and invaded by igneous rocks, syenite, porphyry, basalt and trap. In fact the resemblance to the rocks of the Ural is so striking, that Sir R. Murchison, on examining specimens produced by Strzekecki in 1844, at once predicted that gold would be found there.

In 1851, the experience of California was brought to bear on Australia, by Hargrave, and by June of that year $200,000 worth was produced from the Bathurst diggings. Many persons arrived from all parts, and the immigration from England increased from 2,458 in 1850 to 125,000 in 1852.

The gold discoveries extend at least over nine degrees of latitude, and occupy a breadth of fifty miles or more, *along the line of junction* of the Palæozoic and eruptive rocks. The gold is of great fineness, the different analyses in 1854, only giving from 3 to 7° silver.

The auriferous deposits present the most striking analogy to those of California. There are innumerable diggings, which are sometimes quite superficial, and sometimes extend to the "bottom rock" or rock in place through the detritus and loose materials. In the surface workings which are exclusively in the flanks of the hills, the gold is diffused through the gravelly soil to the depth of a foot, beneath which there is a bed of stiff red clay, which contains little or no gold. In the deeper workings it is sometimes necessary to sink twenty-five or thirty feet, or even more, before reaching the auriferous deposits.

The channels of the small streams, coming down from the mountains, and often entirely dry for the most part of the summer, are usually rich in gold, which is *found accumulated against the "bars"* or projecting ledges of rock, and forced into the chinks between the strata. The more precipitous the torrent, the larger the nuggets which are found ; when the valley expands out, the golden particles are smaller. The rich deposits were frequently found at the *entrance of a lateral into the main channel*, where there has been an eddy in the current, caused by the contraction of the mouth of the lateral valley.

In the valleys of Bendoc and Delegete, the Revd. M. Clarke gives the superficial deposits, as under ;

1. Gold bearing detritus made up of fragments of slate and quartz, cemented by argillaceous matters ;
2. Pipe clay ;
3. Erratic blocks or pebbles of quartz containing gold ;
4. Rock in place.

Any person who has visited the diggings of the "Gilbert" a little above its junction with the Chaudière, must be struck with the close resemblance the above discloses, to our Canadian diggings.

Quartz is very abundant and is the gangue of the gold. Magnetic iron sand, sometimes titaniferous is rarely wanting. Topaz, Garnet, Zircon, &c., are abundant.

The question whether the deposits are the result of gradual and long continued action, which is still in operation, or of a drift period, similar to the "Northern drift," cannot be decided, but M. Scutchbury the Government Geologist and the Revd M. Clark favor the former hypothesis.

Mr. Whitney, writing in 1854, says: "The whole amount of gold hitherto obtained, almost without exception, has been from washing, and though numerous quartz mining Cos. have been formed, they have failed to accomplish any thing." But I find that as early as the following year, symptoms of exhaustion were evident, and general attention was being directed to quartz mining, and on referring to the Prospectus of the Mt. Egerton Quartz Crushing Co. issued in that year I find it stated : "The portion of the vein which the Co. propose working, varies from 12 to 24 feet in width, and like all other veins of a similar nature, *increases in richness the deeper it is wrought*."

It is worthy of note, that at this time, Mr. Hopkins, an English Engineer of well known scientific attainments, gave it as his opinion, that there would be a gradual falling off in the yield of the diggings, and that *there were no quartz veins, worthy of being worked.*

But in 1863, we are able to throw a little more light on the subject. We know that in 1851, the total amount washed, was 30,000 lbs., and that in 1852, owing to the great increase of diggers, it reached 330,000 lbs., but from exhaustion in the diggings, the supply fell off in 1853 to 210,000 lbs.

Attention was then turned to quartz mining, and *reefs*, as the schistose and quartose formations are called in Australia, were discovered and worked by companies. Some of the reefs have proved exceedingly rich ; the Bulleen mine near Melbourne, yields gold, *iron* and *antimony* from every piece of quartz broken ; at certain points, as much as 22 oz. per ton. In a late number of the " Mining Journal " reference is made to about thirty differents Quartz Mining Companies, all in successful operation, though according to official reports, some of the districts do not give more than an average of 2 dwts. 10 grs. of gold per ton, the greatest average being 9 dwts. 16 grs. It is worthy of special mention that the Black Hill Company obtains an average less than 2 dwts. per ton, yet pays 24° on its capital. The Prince of Wales Company has yielded 200 oz. per ton.

At some points the auriferous drifts are of great breadth and depth, and are worked by shafts, down to the schistose and quartose auriferous rock. The black lead is said to resemble the celebrated blue lead of California. Many of the quartz mines are worked below the drift, as in the " Clunes" gold field, and have attained depths from 120 to 500 feet. A commission lately appointed by Government have reported, that they believe the quartz lodes *diminish in value as the depth increases*, but the Mining Journal shews that this opinion has no satisfactory basis, and though some state, that lodes rarely pay below 500 feet, yet Mr. Crichton of the Victoria Company, Clunes, reports that at a depth of 410 feet, *the lodes are as rich as ever*, and Mr. Meurice, of the Clunes Company, gives the following return :—" The average yield of the quartz down to 225 feet, was 5 dwts. 17 grs. per ton ; from 225 feet to 286, the present depth, 8 dwts. 16 grs. per ton." Mr. Bland of the Victoria Company raised from a depth of 300 feet, quartz which yielded 20 oz. per ton. Mr. Bland asserts that the decrease in the average arises now, from the quartz not being selected, but rubbish and all

3*

being sent to the mills. But there need be little fear of exhaustion, when reefs are found, which at 120 feet deep, are sixty feet wide, and yielding 15 to 17 dwts. to the ton.

The result of quartz mining in Australia may be best appreciated from this fact ; the yield of gold which we have seen fall to 210,000 lbs. in 1853, had risen through this agency, to 327,930 lbs. in 1857.

The total value of the gold of Australia from 1851 to 1859 has been estimated at $500,000,000. Fawcett in his " Political Economy," estimates the average annual product, at £4,000,000 for the twelve years ended 31st Dec., 1862. But, (as if to rebuke the shortsightedness of statesmen) the most remarkable feature of this sudden and enormous influx of gold into the world, was its influence on the localities where the discoveries were made.

Along with much of the scum and sediment of society, immense numbers of hardy and energetic and intelligent men were attracted ; their wants gave an immense and healthy stimulus to commerce, settle- ments became towns, towns wealthy and populo cities ; as soon as the lust of gold was satisfied, lands were acquired and the desert was made to blossom as the rose ;" Australia did not fall off as a wool- growing country ; its agricultural productiveness was enormously increased ; it became a grape growing country ; and the wines of Australia promise to rank side by side with the choice vintages of the old world. Education, science and art have been fos- tered and encouraged, and Australia is now looked upon as one of the most prosperous of the colonies of Great Britain.

In concluding this part of our subject, it may be interesting to compare the product of the Californian

and Australian gold fields for the three most mo-
mentous years :

	1851.	1852.	1853.
California....lbs.	262,000	252,000	252,000
Australia..... "	30,000	330,C00	210,000

BRITISH NORTH AMERICA.

Nova Scotia.—In 1855, Dr. Dawson published in
his " Acadian Geology" a description of the " Atlantic Coast Metamorphic district," which is about 250
miles in extent from Cape Sable in the west to Cape
Canseau in the east, and consists of clay slates and
quartz rocks ; the slates of various-degrees of coarse-
ness and of grey and black tints ; the quartz being in
thick massive beds of a grey color, and being com-
monly called " whin." In some localities these are
replaced by mica slate and gneiss, penetrated by veins
and masses of intrusive granite. These rocks are
considered to belong to the *lower part of the lower
silurian* and contain no trace of organism, and are
considered to be the equivalents of the Australian,
Californian and Canadian auriferous rocks. The doc-
tor indicated the probability of these Nova Scotian
rocks being auriferous, but gold was not discovered
till 1860, when it was the result of accident.
  The gold fields of Nova Scotia have hitherto yielded
very little gold from washings, which are confined to
the deposits of sand between high and low water
mark on the Lunenburg coast. The quartz has been
and still promises to be the chief and permanent
source of supply. It is supposed that from the close
vicinage of the gold bearing mountain range, to the
coast, and from the grooves in the rocks indicating the
direction of the glacial drift, all the disintegrated
matter was carried into the Atlantic, and deposited
at its bottom, and a portion of it thrown up with the

sands of Sable Island, in which gold has been found
in appreciable quantities.

On the first discovery, some thousands of persons
flocked to the diggings, and it was expected that in
the following year an immense immigration would
have taken place, but though a very large amount of
gold in connexion with loose quartz, disintegrated
from superficial boulders, was picked up, yet before
the close of the season of 1861, the supply had become so small and uncertain that most of the gold
seekers had left the fields, and had not there been
among their number several returned Californians
and Australians, whose experience enabled them to
form a more correct estimate than the others, the probabilities are that no more, for some time at least,
would have been heard of the Nova Scotian gold
fields. But in 1862 through their instrumentality
several veins were tested successfully and at the close
of that year the Commissioners reported that on an
average 484 men had been engaged in mining during
the season, and that the average earnings of each
man per day, was $1.18.—Compare with this, the
return to the Victoria Legislature in 1860 which
shewed that during that year, 18,296 men had been
employed in quartz crushing, and that the whole product had been 93,025 oz. of gold, giving each miner
an average of 31 cents per day only.

The Nova Scotian Gold Commissioner also reported
that the leads increased in thickness as in depth, but
that it could not be said that they increased in richness. But that when gold was found on the surface,
it was also found equally distributed to the depth
then mined.

We have now the official reports to the close of
1863. We find that nine gold fields have been
established, and twenty-four steam and ten water
crushing mills built ; that the greatest number of

men employed at any one time was 1,156 on 127 areas or claims, and the smallest number 715 men. That the yield of gold was 13,991 oz. of the value of $279,820. The maximum yield per ton was 65 oz. the lowest 4 dwts. 15 grs. During the quarter ended 31st Dec. last, 4,848 tons of quartz had been raised, yielding 4,178 oz. gold, and giving an average of 18 dwts. 1 gr. per ton. The maximum yield in Australia in 1860, was 200 oz. and the lowest under 2 dwts. per ton ; the average being as low as 9 dwts. 16 grs. We have already noted that St. John del Rey mine gave an equal yield of 9 dwts. 9 grs. per ton.

The surface of the gold field is covered with stones and pebbles and boulders of whin of all hues from' white to blue, slaty rocks and large boulders of gray granite with mica. Under those is soil, and undulated hard whin, interspersed with slaty rocks of various degrees of hardness and in some instances resembling sandstone. The gangue of quartz varies in size from a thread to four feet in thickness ; it may be white or stained with iron, or it may be blue as in the Hewitt lead. It may be imbedded in slate or hard whinstone. The gold may be found in visible quantities called sights, or may be invisible and equally diffused—it may not be confined to the gangue of quartz, but may be diffused invisibly through the slate or whin in which it is imbedded—it may also be found in the " dirt" or soil superimposed and adjoining.

The experience of the past year according to generally expressed opinion, has shewn, that as a general rule the lode increases in richness as in depth. And though the Nova Scotian gold fields have not excited the interest which followed the discoveries in Australia and California, the progress made during the past year, warrants the belief that they will go on steadily increasing in productiveness and eventually become one of the permanent branches of Provincial industry and wealth.

The coal mines of Nova Scotia though nearly half a century in operation, have not hitherto given employment to eight hundred miners, though upwards of 100,000 chaldrons have been for some time past annually exported. The gold fields in the second year of their existence have given employment to upwards of 1,100. The gold mines are all easy of access, labour is abundant and cheap, the same may be said of provisions and all the necessaries of life, and the country, like Canada, is healthy, and the population sober and orderly.

CANADA.

In the 2nd volume of the " Transactions of the Literary and Historical Society of Quebec," published in 1830, will be found a paper by Lieutenant Baddely on " Metallic Minerals in Canada," and at page 343, there occurs this paragraph " Gold has not been noticed in Canada, but so near the frontier as to induce me to give the following extract from an American paper, the particulars of which there appears to be no reason to question the accuracy of." He then quotes the " Boston Traveller" which states that gold had been found in Vermont, in the highway, deposited with a quantity of clay sand and gravel, which appeared to have been brought down by a small rivulet from a high hill. The specimen examined by general Field weighed $8\frac{1}{4}$ oz., was of conical form and had firmly adhering to its base a number of small transparent rock crystals.

On the appearance of this, Professor Eaton wrote in Silliman's Journal—" If General Field's specimen of gold found in Newfane, Vermont, be a true native specimen, we may anticipate the discovery of gold *in the talcose state from Georgia to Canada*, along the east side of the Green Mountain Range."

". Gold," adds Baddely, " *in situ* is almost entirely confined to primary rocks ; it has, however, been observed in secondary. But it is in the alluvial deposits that it is more generally found."

Here then we have a prediction in reference to Canada, which ought in justice to place Eaton and Baddely among the prophets ; we will accordingly rank them with Murchison, Dana and Dawson.

According to the Seignior De Lery, gold was first discovered in the Chaudière district in 1818. It was *first announced to the scientific world* by Lieut. Baddely, of the Royal Engineers, in a communication to Professor Silliman, published in the American Journal of Science, vol. xxvii., page 112, in the year 1835, after the discovery of the nugget at the mouth of the " Touffe des Pins," in 1834, by the woman Gilbert.

In 1847, examinations were made by the geological survey, which shewed that gold was not confined to the Chaudière, but existed in superficial deposits over a wide area.

In 1848, Sir Wm. Logan made a report to Lord Cathcart, which was submitted to the Legislature, that in addition to indications by quartz veins, " There is an alluvial auriferous deposit in the Seigniory Rigaud de Vaudreuil, on a small stream called the Touffe des Pins, a tributary falling in on the right bank of the Chaudière, about 58 miles from Quebec—that one or two quartz veins run under this auriferous deposit, and that it was not improbable that in these or other contiguous veins the source of the gold field would be found."

In 1853, Lord Elgin, after visiting the Chaudière, sent a despatch on the subject to the Duke of Newcastle.

Messrs. Thomas Abell & Company's assay, made
at that time, showed the gold to be 3¼ grs. worse
than standard, with 28 dwts. of silver to the pound ;
and in the same year, Messrs. John Taylor & Son
sent out Messrs. Arthur Phillips and Arthur Kent to
survey and report. Their very minute report bears
date, 7th July, 1853, and discloses every thing yet
known of the Des Plant, Famin, &c. The public
were then made aware that the black sand washed
out of alluvial diggings by Mr. James Logan, gave
9 oz., 9 dwts., 18 grs. of gold to the ton.

In the same year, Dr. Sterry Hunt, in page 370, of
the Geological report, said—" a vein which occurs
at the rapids of the Chaudière, in the parish of St.
Francis, Beauce, contains in a gangue of quartz,
galena, blend, arsenical sulphuret of iron, often well
crystallized, besides cubic and magnetic iron pyrites,
and native gold in minute grains. A portion of the
galena from the assorted and washed ore, still containing a mixture of blend and pyrites, gave by assay
69 per cent. of lead, and 32 oz. of silver to the ton, 2,240
lbs. of ore. Another portion more carefully dressed,
gave at the rate of 37 oz. of silver per ton ; and
the button of silver obtained by cupellation from this
lead, contained a small but appreciable quantity of
gold ;" and " a quantity of gold dust from the washing of the sands of Rivière du Loup, was submitted
to amalgamation, and left one third of its weight of
black sand, of which 18 per cent. were separable by
the magnet—the non-magnetic portion dissolved by
the successive action of hydrochloric acid and bisul-
phate of potash, leaving four eighths per cent. of
silicious residue. The solutions contained iron and
chromium gave by prolonged ebullition 23.15 per
cent. of bititanic acid."

In 1851-2, the Canada Gold Mining Company
made an experiment on the River du Loup, at its

junction with the Chaudière, and it is on record that
" the gravel from about ⅔ of an acre, with an average
thickness of two feet, was washed during the summer
of 1851, and yielded 2,107 dwts. of gold—value
$1,826 ; profit $182. Price of labour, 60 cents per
man per day.

In 1852, ⅗ of an acre was washed, giving 2,880
dwts. of gold=$2,496 ; profit $1,366 ; and the fineness of the samples 864.000, 871.00ʋ and 892.000.
" And there is no reason," says Dr. Hunt, " for sup-
posing that the proportions of the precious metal to
be found along the St. Francis, the Etchemin, and
their various tributaries, is less considerable than that
of the Chaudière."

Dr. Hunt's recent brochure, published in advance
of the Geological report, has furnished reliable data
down to the present time, which have been fully corro-
borated by practical experience on the Gilbert during
the past season. But though little disposed to under-
rate the value of the " placers," we are inclined to
attach the greatest importance to the following
announcement by Dr. Hunt : " The precious metal
occurs again not far from Harvey Hill Copper Mine
in Leeds, at a locality known as Nutbrown's shaft,
which is sunk on a vein of bitter spar, holding specular iron ore, vitreous copper ore, and native gold
generally in small grains or scales ; *some specimens
from this locality weighed as much as a pennyweight.*"

Gold in quartz rock has also been found at the
" Devil's Rapids," on the Chaudière, and a report is
abroad that the Handkerchief Peak in St. Sylvester, has
furnished from its quartz veins very rich samples of
gold with gray and purple copper ore. Having per-
sonally inspected several of the Nova Scotian gold
fields, and compared their natural aspect with the
Chaudière region, we have never since doubted that
gold would be found in abundance in many of its

quartz rocks, which must eventually be the main source of supply in Canada, as in all other countries. We do not pretend to such an intimate knowledge of the gold regions of Canada as we would desire, but from what we have seen, coupled with what we have read, and what we have learnt orally from a gentleman most thoroughly acquainted with the subject, we would hesitate to express any decided opinion as to the extent to which the placers may be profitably worked ; but from the descriptions of the deposits in California and Australia, and with the statement made by Dr. Hunt, we are inclined to believe that operations prudently directed during the coming season, may discover deposits not inferior to those far-famed placers.

Dr. Hunt has pointed out, that according to Blake, a bushel of earth containing *one twenty-fifth grain* of gold, has proved sufficiently remunerative, under the hydraulic system of California. We have it on the best authority that the sands of the Ural are worked with profit, though yielding only *three fourths of a grain* per bushel of 100 lbs. And we have on the authority of Dr. Hunt, that the drifts of the Chaudière contain *one and three fourth grains* of gold per bushel.

We must also bear in mind, that Dr. Hunt has also assured us, that gold is not confined to the gravel of the river channels and alluvial flats, and that the banks of the Metgermet *of interstratified clay and gravel* (similar to those of Australia and California) were found to contain gold through their whole thickness of fifty feet, and that the best conditions for the application of hydraulics, present themselves on all the tributaries of the Chaudière.

We should not like, however, to be understood to say, that every adventurer will make his fortune ; we would rather discourage those who are at present engaged in useful and profitable occupations, from following the " ignis fatuus '' of the gold seeker. The past season may furnish a profitable lesson ; a few like the Poulins will realize handsome returns, a few more may make fair wages, the vast majority will be disappointed. An experienced Australian miner, who has visited the district, affirmed to me that where so much " heavy " gold has been found, much " *fine* " gold will be found ; this we believe, but it will be a lottery, and who will venture to say, I will draw a prize. " Chemists, Geologists and Mineralogists," says Hittel, " have not done better than ignorant men and new comers. Most of the best veins have been discovered by poor and ignorant men. Not one has been found by a man of high education as a miner or geologist. No doubt geological knowledge is valuable to a miner, and it should assist him in prospecting ; but it has never yet enabled anybody to find a valuable claim."

For the benefit of those who have not visited the Chaudière, we will now give a short description. The river extends from the St. Lawrence to the Lake Megantic, in the State of Maine ; its course exceeds a hundred miles. The valley is very beautiful, moderately fertile and healthy. It is inhabited almost entirely by French Canadians. There is a very good road running up the valley, which is daily travelled by the mail courier. About every nine miles there is a pretty village, generally the centre of the parish. Here, and at comfortable inns at other points on the road, the traveller can get accommodation on moderate terms. The population of the valley exceeds fifteen thousand.

The Seigniors hold their patents from Louis Quatorze, which generally bear date 1736, and all require that the Seigniors " shall give notice to His Majesty, or to us and our successors, of the mines, ores and minerals, which may be found within the said extent of land."

ROYAL MINES.

According to the law of England, all mines of gold and silver are Royal mines, and they are so peculiarly a branch of the Royal Prerogative, that it has been said, that though the King grant lands and all mines in them, yet Royal mines will not pass by so general a description.—Plowden, 336.

It had long been decided that all gold and silver mines within the realm, though in hands of subjects, belong exclusively to the Crown by prerogative, and that this right is also accompanied with full liberty to dig and carry away the ores, and with all such other incidents as are necessary for getting them.— Queen v. Earl of Northumberland—Plowden, 310, 336.

The right of entry was disputed by Lord Hardwick, in the case of Lyddal v. Weston—2 Atk. 26—when there was a grant from the Crown of lands, with a reservation of all Royal mines, *but not of a right of entry*, on the ground that by the terms of the grant, there was no such power in the Crown, and that by the Royal Prerogative of mines, the Crown had even no such power, for it would be very prejudicial if the Crown would enter into a subject's lands, or grant a license to work the mines—but that when they were once opened, it could restrain the owner of the soil from working them, and could either work them itself, or grant a license to others to work them.

But in Seaman & Vaudrey, 16 vs. Sir W. Grant, the Master of the Rolls, declared this doctrine *to be liable to considerable doubt*, as being inconsistent with the ruling of the twelve Judges, in the case afore cited, of the Queen v. Earl of Northumberland, in the reign of Elizabeth, and " Bainbridge " says " it may therefore be assumed, that the latter case which was solemnly decided by all the twelve Judges, has never been overruled."

It has been decided in Earl of Cardigan vs. Armitage, that the right to enter and work mines, *is necessarily* incident to a grant of mines, without any express authority for that purpose, and that this power cannot be restained by a special power given in the *affirmative*, which may authorize more acts than would be implied by law, but which will in no way exclude the full operation of the law.

This case established the right of the Crown to enter the lands of a subject, to search for and work mines, and also to convey that right to others, and further that the *owner of the mines*, when they formed a distinct inheritance and possession, was entitled to work them, *without the concurrence of the owner of the surface.*

The Canadian patents which we have seen, all contain the reservation of all gold and silver mines, and also express right of entry, and to use and make roads, and to divert rivers.

Under the French crown, the civil constitution of Canada was established on the feudal system, and at the time of the introduction of the Custom of Paris, previous to subinfeudation, both the *dominium directum* and *dominium utile*, were in full demesne in the seignior. Previous to the arrêt of 6th July, 1711, known as the edict of Marly, the concession of lands to *habitants* was not obligatory on the seignior, but it was then made so. The effect of the edict was to divide the estate between the seignior of the fief, and his subfeudatory or *censitaire*, in such manner as to retain to the former the immediate demesne, and to convey the useful demesne to the latter. The immediate demesne consisted of the duties or dues, obligations or redevances, to which the *censitaire* was subjected, such as *foi et homage*, and the *cens et lods* imposed by the seignior. The useful demesne consisted in the product of the soil, which the *censitaire* had the

4

right of occupying as proprietor, and comprised the use of the unnavigable waters and forests connected therewith. Whenever the seignior conceded lands to *habitants*, he had no right to make any reservations, and the *habitant* became proprietor of the soil which he occupied, subject only to the duties above mentioned. This law remained in full force at the cession of Canada by France to England, after which all concessions under the French crown were confirmed by ordinance issued from the castle of St. Louis, in 1766, by Governor Guy Carleton, and continued in force at the passing of the seigniorial act in 1854, which is now the law of the land.

We believe it is admitted that the Royal Prerogative, as recognised by the law of England, overrides all merely municipal laws of Canada.

There having been no concessions of mines, (which is tantamount to express reservation,) the rights of the crown in this Province remain intact, unless where affected by the DeLery patent and the act incorporating the " St. Lawrence Mining Company." Nothing therefore stands now in the way of legislation but these and the royal pleasure.

Nothing need be said on the DeLery patent, as it is understood its validity will be tested by the courts. The St. Lawrence Mining Company Act, though conferring very valuable privileges, is not considered to cover either gold or silver mines. The experience of the other colonies has shown that the royal pleasure will not stand in the way of necessary legislation.

This then is the position of Canada as regards this important question, at the present moment. Let us now look at the steps taken in other countries, on the discovery of gold.

The progressive stages have been alike. First, gold washing from alluvials, and secondly, the discovery and crushing of quartz rock ; each requiring the establishment of appropriate regulations.

In Brazil, the mining code of 1618 was severe, but its gradual relaxation, and granting mining ground by the Crown, with considerable power to the miners to regulate their own affairs, became inevitable, from the very nature of their pursuit. At the present day the code is most liberal : even the royalty being waived.

In California, as in Brazil, the code has been gradually ameliorated, all injudicious attempts of the Government to impose arbitrary and impracticable regulations, having signally failed. The question of the right of gold-miners, to search and work for gold on private lands, has been settled the same way as in Brazil, the gold as treasure trove belonged to the finders, who were only bound to pay the ordinary royalty, and to indemnify the proprietors for any loss in crops, or other damage the land might sustain.

But by a recent decision of the United States Court, Colonel Fremont has successfully asserted his exclusive rights as proprietor, to all the mines on the Mariposa estate, in extent seventy square miles, and a company, with a capital of $10,000,000, has been formed to work them.

In Australia in 1851, the Government issued its proclamation, asserting the right of the Crown to the newly discovered gold fields, and established temporary regulations under local gold commissions and subsequently in 1853, promulgated a complete code.

In alluvial washings, 1st claims of twenty feet frontage on either side of a river, were allowed to each miner ; 2nd, twenty feet of a tributary to a river extending across its whole breadth ; or 3rd, sixty feet of the bed of a ravine or water-course ; or 4th, twenty feet square of table land, or river flats. A license fee of thirty shillings a month to be paid by each miner. Larger auriferous tracks to be conceded to companies on special leases.

The quartz matrix claims, to consist of half a mile each in the course of the vein, with an average breadth of a quarter of a mile on each side, reserved for building and other necessary purposes. Application fees of £50 for each quartz matrix claim, were to be paid in the first year ; and the royalty, instead of a monthly license fee, to be ten per cent. on all gold produced. The duration of the lease to be twenty-one years, and surety to be given for the due payment of royalty, and twenty persons, or an equivalent in horses and machinery to be employed within twelve months. These regulations were either in themselves so intolerable to the class of persons who flocked to the diggings, or were so injudiciously enforced, as to lead to general disaffection, and armed resistance to the authorities, terminating on 3rd Dec., 1854, in collision between the miners and military and police, at Ballarat, when twenty of the miners were killed and about sixty wounded.

A commission of enquiry considered that the malcontents were to some extent justified. The royalty was reduced to three per cent., and the licence fee to a payment of £1 per annum. Subsequent acts of the local parliaments have authorized the enactment of regulations, by the local mining boards at the various gold diggings ; and these have all the force of law on receiving the subsequent approval of the Governor. The local regulations consequently now differ from each other materially according to the nature of the gold deposits, in each district. The imperial government has also conceded the right of the mint. The export duty, which was, at first, two shillings and six pence sterling per oz., is still enforced at the reduced rate of two shillings per oz.

Before concluding the remarks on Australia, it may be mentioned, that quartz mining has to a large extent superseded alluvial diggings ; to carry which out, costly machinery is required. The necessity for capital is thus created; and capital must be looked for outside the working miners. Before capital can be invested, a certain fixity of tenure must be established. The Government directed their attention to this feature, and an act of the Victoria Legislature, passed in 1858, authorized the Governor in Council to issue regulations on the subject of partnerships for mining pursuits. The miners as a body were disinclined to accept the condition of hired servants, whilst the capitalist on the other hand would find his investment fruitless, without the assistance of the experience and skill of the practical miner. The Government, after much deliberation, at length issued a model form of instrument for associations. According to its provisions, mining associations may comprise, property partners, labour partners, and money partners. The model instrument consists of numerous clauses, of which each company may adopt as may it pleases, provided that no other conditions are adopted inconsistent with its general scope.

Last but not least, comes Nova Scotia. Immediately on the discovery of gold in Nova Scotia, in 1860, the Government asserted its right (the right of the Crown), to the gold, and issued temporary regulations, which remained in force till the passage of the Act of March 1862 ; under these regulations, an officer was sent, who took possession of the field in name of the Government, laid off claims and made arrangements for the maintenance of peace and order.

Under the Act of 1862, a chief gold commissioner and deputies were appointed, who, within the limits of the several gold fields, exercised the powers of Justices of the Peace ; districts were surveyed and declared gold fields, according to the meaning of the Act, the Government resuming such portion of the field, if any hand been granted, the owner receiv-
4·

ing the assessed value of the land, without reference to its enhancement in value from the discovery of gold, and twenty-five per cent bonus. Quartz mines were laid off, in areas of different sizes. Area No. 1, being one hundred and fifty feet along a lead, by two hundred and fifty across. Rent forty dollars per annum.

Area No. 4, being four hundred and fifty feet along by five hundred across. Rent two hundred and forty dollars per annum.

Under special circumstances, leases of larger areas, and on modified terms, might be granted with the approval of the Governor in Council.

Alluvial diggings were to be laid off in lots of one thousand square feet. Rent five dollars per annum, payable in advance.

A royalty of three per cent to be exacted upon the gross amount of the gold mined.

Under this Act, several gold fields were gazetted, claims were leased, shafts sunk, and order maintained among the miners during the season of 1862, and considerable progress was made in developing the true system of mining. But the experience of the year also elicited the defects of the law, and led to the passing of the Act of March 1863, under which, any number of areas of any of the classes prescribed by the former Act, not exceeding five, if lying contiguous to each other, might be included in one lease, and the labour to be put on the demised premises, being at the rate of one hundred days, for each area of class No. 1, might be put on any part of the demised premises.

Applicants for leases of mining areas on private lands, without the limit of any proclaimed gold district, or on any lands within the limits of a proclaimed district, not ordered under the principal Act to be revested, may arrange by agreement in writing with the proprietors for leave to enter, and for easements and damage to lands. In such cases, instead of paying at the rate of ten dollars per annum per area of class No. 1, there shall only be a payment at the rate of two dollars.

The chief gold Commissioner may issue licenses to search for gold, to be called " Prospecting Licenses," to include an area not exceeding twenty-five acres, and not to be in force for a period exceeding three months. But the written consent of the proprietors (in case of private property) must accompany the application for prospecting license.

Mills erected for the crushing or reduction of quartz, must be licensed. And the owner or occupiers must enter into bonds, in the penalty of two thousand dollars, to keep a book of account (to be supplied by the gold Commissioner) in which shall be entered a clear and distinct statement of all quartz crushed with the name of the owners; weight of each lot, date of crushing, actual yield in weight of gold, the royalty at three per cent, and lastly, the shaft or area from which the quartz was raised. Three per cent to be retained as royalty and paid over weekly to the gold Commissioner, or its equivalent in money, at the rate of nineteen dollars and fifty cents per oz. troy for smelted gold, and eighteen dollars and fifty cents for unsmelted gold.

Under these regulations, gold mining has made rapid but healthy progress. Twenty-one steam and eight water mill crushers have been erected. On the average one thousand men have been engaged in mining during 1863. In July the maximum yield to the ton of quartz was thirty ounces of gold; in August fifty-six ounces ; in September sixty-six ounces. During the quarter ended September 30th, on an average nine hundred and sixty-six men were engaged, who raised 4,620 ounces, 2 dw. 2 gr. of gold, equivalent to $88.45 per man, for seventy-nine working days.

There has been no collision between the mining population and the authorities, and order has been maintained without difficulty among the miners themselves.

The details in connexion with Nova Scotia possess the more interest for Canada, from the fact that the Chaudière gold fields in many respects closely resemble those of Nova Scotia. We do not say that the Nova Scotia law is altogether adapted to the condition of things here, but connected with the facts furnished by Brazil, California and Australia, it may furnish some useful hints, of which the Government of Canada may make good use in their proposed legislation.

Our readers who have travelled with us over the several gold fields, must have been struck with the close resemblance in their natural and political history. Each country at first looked to its superficial deposits for its supply of gold, and each in succession, on their exhaustion, either ceased to be gold producing, or as in the recent instances of Brazil, California and Australia, whence a correct knowledge of the origin of gold had been acquired from " stubborn facts," each turned its attention to the development of the wealth concealed in its auriferous rocks.

Their united experience proves that though gold may not be got with the same ease, from the deep mine as from the surface placers, the quantity produced is at least equal in amount, and promises to be permanent.

We find gold and iron are almost as intimately connected, as gold and silver, which in nature are inseparable, and Whitney pointed out in 1854, that there is hardly a specimen of iron pyrites, in which a minute trace of gold may not be detected ; and further, in

1858 it was discovered in Colorado, that the iron pyrites (Fools Gold) *was very rich in fine gold.* This ought to be particularly borne in mind, as quartz veins charged with iron pyrites, are reported to abound in our Canadian gold regions, where also is found quartz in combination with iron and copper pyrites.

We must have observed in Australia and California, deposits of interstratified clay and gravel, like those on the Metgermet which have been proved auriferous.

We have remarked in almost every gold region, deposits of black sand containing Titania and Garnets and Gold, which are the equivalents of those of the Chaudière.

In fact the Canada gold region combines most of the favorable indications in the gold fields which have passed under our review. The Chaudière offers great facilities for testing the extent of its metallic wealth ; the abundant water power can be economically brought into play, against the diluvial deposits, and we have the experience of California to guide us in the application of hydraulics. Water power will also in many places operate on mills for crushing quartz. There is also abundance of wood for roasting ores, a process which in Australia and California, has been found not only to facilitate the crushing quartz, but by driving off the sulphur and arsenic, renders amalgamation with mercury easier and cheaper.

There may not be any great rush of immigrants to the Chaudière on the opening of spring ; we hope that there may not. But as we know that gold washing can be prosecuted with little capital, and with the risk only of the loss of a certain amount of time and labour, we trust that the Canadian Government will encourage by a liberal enactment, the enterprise of individuals, and at the same time give all reasonable facilities to capitalists and associations.

We have seen in every case, where vexatious and onerous terms were imposed, mineral enterprise languish and die, but as soon as these were relaxed, and a liberal policy inaugurated, the products of the mines rapidly increase, commerce stimulated, population increase, and general prosperity ensue.

In conclusion we have much pleasure in acknowledging our indebtedness for much of the information contained in this *compilation* to the works of Murchison, Whitney and Ure, Logan and Dawson and to the very interesting brochures of Dr. Hunt and the Revd. Mr. Douglas ; we have also quoted largely from Hittel. We concur in opinion with Mr. Douglas whose intimate acquaintance with all past operations on the Chaudière entitles him to speak, that Canada will not likely turn out an Australia or California, but we do believe that by judicious management, it will gradually but surely, somewhat like Nova Scotia, take its rank as an important gold producing country.

NOTE.—The gold fields of British Columbia are reported of great extent and richness both in alluvials and quartz rocks, and do not differ in any material point from the other great gold regions.

The gold fields of New Zealand are of great importance, and are "placer diggings" only, and Sir R. Murchison has informed the Royal Geographical Society, that the summits of the adjoining ridges are not now denuded, *but covered to the depth of many feet with soil.*

www.ingramcontent.com/pod-product-compliance
Lightning Source LLC
Chambersburg PA
CBHW022029190326
41519CB00010B/1640